TIGERS

TIGERS

PAULA HAMMOND

amber
BOOKS

This pocket edition first published in 2023

First published in a hardback edition in 2020

Published by
Amber Books Ltd
United House
London N7 9DP
United Kingdom
www.amberbooks.co.uk
Instagram: amberbooksltd
Pinterest: amberbooksltd
Twitter: @amberbooks

ISBN: 978-1-83886-260-2

Project Editor: Sarah Uttridge
Designer: Jerry Williams
Picture Research: Terry Forshaw

Printed in China

Contents

Introduction

"Tyger Tyger, burning bright,
In the forests of the night;
What immortal hand or eye,
Could frame thy fearful symmetry?"

WHEN THE ENGLISH POET William Blake wrote this in 1794, few Europeans had seen a tiger. They only knew of this exquisite creature through wild tales told by visitors to Asia and the Far East. Blake admired the tiger both for its beauty and ferocity. Other people have worshipped it, immortalized it in legend, and adopted it as their national emblem. Northern India's Naga tribes even believe that people and tigers are kin – one human, the other feline.

Sadly, many don't share these views. For hundreds of years, hunters – fuelled by fear or the greed for trophies – have driven this magnificent creature to the brink of extinction. With tiger habitats shrinking, and illegal trade in tiger parts ongoing, this big cat's future looks dire. So dire that most of the images in this book were taken in nature reserves and zoos.

This volume is a celebration of the tiger. It's also a glimpse of a world in peril – but it's a world that we can still save if we have the will to do so.

OPPOSITE:
A playful, juvenile Bengal tiger splashes through the water.

Species

Tigers (*Panthera tigris*) diverged from other members of the genus *Panthera* during the Ice Age, when this once widespread mammal shared the Earth with mammoths, mastodons and early humans. While some groups, such as the Japanese tiger, died out, others flourished, and all of today's tigers are believed to be descended from a common ancestor who lived around 72,000–108,000 years ago.

Tigers are the world's biggest wild cat and the Siberian tiger is the biggest of them all, averaging up to 3.3m (11ft) from head to rump, with another metre (3.2ft) for the tail. In comparison, Sumatran tigers are the smallest members of the group, growing up to 2.4m (7.9ft). While Siberian tigers may be the biggest, Bengal tigers are the heaviest, weighing in at 325kg (717lb), which is twice the weight of a refrigerator.

Despite these variations in size and, occasionally, colouring, the world's tiger populations are genetically very similar. Zoologists have long argued about how many sub-species there are. In 2018, the geneticist Yue-Chen Liu studied the genomes of 32 specimens and identified nine distinct sub-species. Sadly the Bali, Caspian and Javan tiger are all extinct. Five sub-species – the Bengal, Indochinese, Malayan, Siberian, and Sumatran – still survive in the wild.

OPPOSITE:
Siberian tiger
Many big cats have spotted coats which break up their outline while hunting, but the tiger's most distinctive attribute is also its most beautiful. Tiger stripes vary between sub-species but are usually dark brown or black on an orange-brown background. Like fingerprints, every tiger's patterning is unique.

PREVIOUS PAGES:
Cool cats
Tigers may be primarily nocturnal hunters, but by day they can often be found lounging near rivers and streams, taking regular dips to stay cool. A strong body and webbed paws make tigers excellent swimmers and they're easily fast enough to pursue and catch prey in water.

OPPOSITE:
Tree-huggers
Tigers scratch trees to keep their claws sharp and clean. Between their claws they have inter-digital glands which release a scent when the cat scratches the bark. This scent-marking is used to claim territory, and make other cats aware of their presence – avoiding potential conflicts.

ABOVE:
Dressed for the weather
These more northerly Siberian tigers have longer, thicker and paler orange coats than their more southerly cousins. Beneath a tiger's stripy coat is an additional layer of fluffy fur which helps to keep it warm. Amazingly it's not just a tiger's fur that's striped. Its skin is too.

PREVIOUS PAGE LEFT:
Roaming free
While lions are creatures of the open savannah, tigers have adapted to live in a range of environments, from damp, sub-tropical jungles, to swamplands, and upland forests. Tigers still roam wild in 13 countries.

PREVIOUS PAGE RIGHT:
A real mouthful
A healthy tiger has 30 teeth. Their huge upper canines are the largest of all the big cats and can grow up to 7.6cm (3in) long. These are the teeth used to kill. The gaps between the flatter molars and canines makes it easier to grip prey.

RIGHT:
The cat's whiskers
Tigers have five types of whiskers: superciliary whiskers on their eyebrows, genal whiskers on their cheeks, mystacial whiskers on either side of their muzzles, carpal whiskers on the back of their front legs and tylotrich whiskers throughout the body. These are tipped with sensitive proprioceptors.

Bengal tiger

Naturally, tigers occupy a wide variety of habitats. Siberian tigers can be found in sub-arctic forests, where night-time temperatures fall to -40°C (-40°F). In the mangrove swamps of the Bay of Bengal, Bengal tigers tolerate temperatures of 40°C (104°F), bathing regularly to cool down.

ABOVE:
Play fights
A pair of Bengal tiger cubs engage in a little play fight, in preparation for adulthood. Bengal tigers reach sexual maturity at around four or five years of age. In the wild, their life expectancy can be as little as ten years. In captivity, they may live up to 25 years.

OPPOSITE TOP:
Body language
Looking threatening or conciliatory can prevent or speedily end a conflict before it gets too bloody. For tigers, wide-open eyes, a mouth slightly agape, and a furiously lashing tail are signs of aggression. A more defensive posture includes flattened ears, bared teeth and narrowed eyes.

OPPOSITE BOTTOM:
Lying on their backs
In the same way as a household pet will show that it's feeling relaxed or playful, tigers will often lie on their backs and show their belly. A subordinate tiger will also yield to its opponent by rolling onto its back in this way.

Climbing trees
Tiger cubs will happily
chase prey into the
tree-tops but the larger,
heavier adults can't pull
themselves up with their
claws in the same way
that leopards do. Instead,
they use speed and power
to propel themselves
upwards and, because
their claws are curved,
they have to climb down
backwards.

LEFT:
On patrol
The size of a tiger's
territories depends
on the region and the
availability of food, but
it may patrol a vast area
– in India, up to 1000km^2
(386mi^2). Being able to
swim is a huge advantage
and some tigers swim
up to 29km (18 miles)
a day patrolling their
territories.

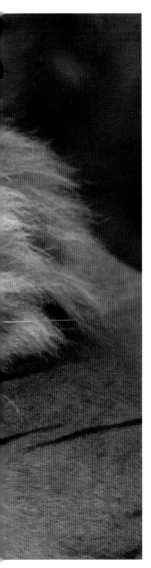

LEFT:

Indochinese tiger

A tiger's tongue is covered in hundreds of small, sharp, backwards-facing barbs called papillae. It's these papillae which give the tongue such a rough texture – able to strip the skin, fur, or feathers from prey. Like domestic cats, they also use their tongues to groom and clean their fur.

ABOVE:

Lapping it up

The tongue is a useful tool and tigers use it to lap up water in the same way as household cats do – only more slowly. It's a balancing act between gravity and inertia as it extends its tongue straight down, with the tip curled, so that the top touches the liquid first.

PREVIOUS PAGES:
Striking stripes
Indochinese tigers can be found in Myanmar, Thailand and Laos. In terms of colouration these rare cats have a darker base coat, with fewer, shorter and narrower stripes. Sadly, due to their rapidly falling numbers, Indochinese tigers are known to inbreed with family members, causing physical deformities.

OPPOSITE:
Chilling out
Tigers can often be found in areas where rivers meet the land – known as the riparian zone. These natural wildlife corridors are where prey meets hunter – sometimes in a surprisingly relaxed environment. Bushes and trees offer shade while water attracts thirsty animals and gives everyone an opportunity to relax and cool down.

ABOVE:
Growing up
At around 11 months old, these Indochinese tigers already look fully grown. As they reach adulthood, young females will start to establish territories close to their mothers'. Males will travel further afield and, if they can't find an unclaimed territory, they may have to challenge another male for his.

ABOVE:
Malayan tiger
Malayan tigers are most
at home in the broadleaf
forest but they will range
far and wide in search of
a meal. Boar, deer and
the goatish serow are
their preferred prey but,
like all tigers, they are
opportunistic hunters
and have been know to
kill sun bears and even
elephant calves.

OPPOSITE:
Wading in
While tigers are much
fonder of water than
domestic cats, swimming
tigers usually only wade
into water up to neck
depth. They'll submerge
their bodies but rarely go
completely underwater as
they dislike getting their
eyes wet. They may even
enter a river backwards to
avoid splashes.

OPPOSITE:
Keen eyesight
All predators have
binocular vision. This
enables them to judge
distances more accurately
than prey species, whose
eyes are positioned on the
side of the head. Tigers'
eyes contain mainly rod
receptors, which are
sensitive to low light. So
sensitive, that their night
vision is about six times
better than humans'.

LEFT:
Paws and claws
A set of large padded
paws help the tiger to
stalk in silence. At
the end of each digit
are claws – up to 10cm
(3.9in) in length –curved
to better grip on to prey.
When not in use, these
formidable weapons
are pulled back into a
protective sheath.

FOLLOWING PAGES:
Sleepy kitty
In a life-or-death moment,
tigers are every inch the
hunter: all speed and raw
power. But catching prey
expends a lot of energy
and, after a kill, a tiger
will gorge on a meal, then
rest, sleeping for up to 20
hours every day.

South China tiger

The South China tiger is the smallest tiger subspecies in mainland Asia. Slightly larger and heavier than the Sumatran tiger, a South China male measures around 2.5m (8ft) in length. Their coats are a beautiful burnt-orange, with a whiter stomach and heavier stripes than their Bengal counterparts.

OPPOSITE:

Walk on the wild side
Each pair of tiger legs is
adapted for a different
type of work. The hind
legs are longer than the
front legs. This gives
the tiger a boost when
it jumps, allowing it to
cover distances up to 10m
(32.5ft). Its front legs are
heavy and muscular, to
pin down struggling prey.

ABOVE:

African tigers?
Historically, tigers
roamed over much of
western and central
Asia – each sub-species
adapting to suit its
environment. Tigers are
not indigenous to Africa
but the male South China
tiger shown here is in
Philippolis, South Africa,
where animals bred in
captivity are being taught
how to be tigers again.

FOLLOWING PAGES:

Home alone?
Tigers are solitary
hunters and well
equipped to bring down
prey such as antelope on
their own. However, tiger
relationships are complex.
Although they're loners,
the males are less
aggressively territorial
than other big cats and
they will tolerate other
males on their patch,
sometimes even sharing
their kills.

41

RIGHT:

Sumatran tiger

The Sumatran tiger became isolated from other tiger populations in mainland Asia when sea levels rose around 6000–12,000–years ago. While increasingly rare, small groups are still found in isolated areas throughout the island and they have been spotted in both lowland and mountain forests.

FOLLOWING PAGES:

Spots and stripes

These female Sumatran tigers can be identified thanks to their stripes. Look closely and they seem to dissolve into scattered spots at the tips. More small, dark spots on the tiger's back, flanks, and hind legs can be seen between the heavy, sometimes double rows of tiger stripes.

Eyes and ears

The white spots on the ears of this Sumatran tiger are common to all sub-species. It's likely that they have a duel purpose. Firstly, to deter attacks from the rear, as they may look like eyes to potential predators. Secondly, as a threat display, as defensive tigers will flatten their ears.

OPPOSITE:

Hear me roar

Domestic cats purr or mew, but only members of the *Panthera* family roar. That's because in place of the epihyal bone, which is present in the voice box of small cats, leopards, lions, jaguars and tigers, all have a flexible ligament. This can be stretched, allowing for a wider and deeper range of sounds.

ABOVE:

Cat climbers

Leopards are the strongest climbers of all the big cats and are known for their habit of hauling prey up trees to stop other predators from stealing their food. Tigers, in comparison, are poor climbers but will climb trees in pursuit of prey or when they are stressed.

FOLLOWING PAGES:

Nose colour

For a tiger there's one tell-tale sign of age: its nose. Young tigers have a pink nose. This gradually darkens, turning orange, then a blackish-brown as it reaches maturity. A tiger's sense of smell is not as acute as its other senses and it's used less in hunting than its spectacular eyesight.

White tiger

A white Bengal tiger owes its striking coat to the lack of red and yellow pheomelanin pigments, which are responsible for a tiger's more usual orange colouration. Naturally, only one in every 10,000 tigers are born white. Snowy tigers have an additional genetic mutation which removes almost all of their stripes.

OPPOSITE:

Growing up fast

Only Bengal tigers carry the gene mutation which produces the characteristic white fur. On average, white tigers are bigger and heavier than their orange counterparts and they appear to mature faster too. White male tigers are considered full-grown after two or three years, while Bengal males don't mature until they're four or five.

ABOVE:

Meeting demand

White tigers aren't albinos. Nor are they a separate sub-species. They're the result of a mutated gene. Their dramatic appearance – blue eyes, white fur and chocolate stripes – has made them popular with theme parks. To meet demand, most white tigers in captivity are the result of deliberate inbreeding, which leads to physical deformities.

FOLLOWING PAGES:

Curious coats

Many Arctic animals grow snowy white coats, which act as additional camouflage in the winter months. However, when the temperatures start to drop, a white tiger's coat will begin to grow in a darker shade. This curious attribute is due to enzymes in their fur which respond to heat.

ABOVE:
Taking a dip
Tigers are less comfortable in the heat than most big cats. Here, a white tiger plunges into a river in a Singapore nature reserve to cool down. He may spend up to an hour at a time in the water before relaxing on the river bank. He'll take regular dips throughout the day.

OPPOSITE:
Courting controversy
Animals in captivity often behave – or are forced to behave – in ways which would be unusual in the wild. This white tiger is shown diving for a piece of meat thrown by his trainers in the theme park where he's kept. Making animals perform for visitors is an increasingly controversial practice.

Tiger Habitats

Once, tigers roamed from the forested foothills of the Himalayas, to the jungles of Bali; from eastern Turkey, to the Sea of Japan. Adaptable and tenacious, these iconic creatures are equally at home in evergreen forests, mangrove swamps, rocky mountains, snowy highlands, coniferous woodlands and grasslands. Like all animals, they thrive where the threats are low and the food, water and shelter are plentiful.

At the start of the twentieth century there were around 100,000 of these charismatic cats. Now there are fewer than 4000. Apart from hunting, the main reason for this is habitat loss. Since the new millennium, tiger habitats have shrunk by 93 per cent. As mankind eats up land for industry and agriculture, these beautiful beasts are increasingly finding that they have nowhere left to go.

India is now home to two thirds of the world's remaining tigers. Twelve other 'tiger range' countries have committed to help preserve and re-establish tiger habitats, including Bangladesh, Bhutan, Cambodia, China, Laos, Indonesia, Malaysia, Myanmar, Nepal, Russia, Thailand and Vietnam. The decline of these iconic big cats is, however, symptomatic of an even bigger problem. The Earth's green spaces are vanishing. We're living in an era of mass extinctions, with the worst species die-offs since the loss of the dinosaurs 65 million years ago. The truth is that tigers face a bleak future unless their traditional habitats can be saved.

OPPOSITE:
Tiger feet
A tiger has five claws on each of its forefeet, including a smaller dew claw which sits further back on the paw. Its hind feet have four claws. Dew claws act like a thumb, and help the tiger to climb or grasp and hold on to prey.

PREVIOUS PAGES:

Heading north

Russia's Ussurisky Nature Reserve was established to protect one of the region's last remaining deciduous conifer woodlands. Sitting on the edge of the Sikhote-Alin Mountains, near the Komarovka River, it's also tiger country. It's in this wooded, mountainous terrain that some of the world's most northerly Siberian tigers can be found.

LEFT:

Healthy habitats

The Bandhavgarh National Park is home to one of India's largest Bengal tiger populations. The reserve boasts wide valleys and streams, with native sal trees on the lower reaches, and deciduous forests on the higher, drier slopes. Such rich biodiversity provides a healthy habitat for both tigers and prey species.

OPPOSITE:

Taiga tigers

Siberian tigers are adapted to live in the cool, coniferous forests of the north. The name traditionally given to this type of biome is the taiga. The taiga covers huge swathes of north America, northern Europe, and the more northerly parts of Japan and Mongolia. Sadly, Siberian tiger populations are mainly now only found in Russia.

ABOVE:

Supersize me

Most of Malaya's tiger population inhabit an area about the size of England (around 130,395 km² or 50,345.79 mi²) on the Malay Peninsula. Poaching and land development means that there's a scarcity of prey species and, as a result, Malayan tigers need super-sized territories to survive.

RIGHT:
Family troubles
Tigers can give birth to
between two and four
cubs, every couple of
years. However, infant
mortality rates are
high. About half of
all cubs don't survive
to adulthood. Even
in protected habitats,
poaching is rife, and new
mothers face increasing
challenges when it comes
to finding a suitable place
to raise a family.

FOLLOWING PAGES:
Far from home
As habitats vanish, it's
a sad truth that there
are more tigers captive
in the USA than there
are in the wild. Not
all zoos have captive
breeding programmes
but, where they do, they
can play a critical role in
tiger conservation. This
Sumatran tiger cub was
born at Flamingo Land
Zoo, Yorkshire, England.

ABOVE:
Safe-havens
Northern India's
Ranthambore National
Park has been a relative
safe-haven for the
country's tigers for
over 40 years. The
Park's terrain alternates
between lush meadows
and dry, deciduous forest
which gives the resident
tigers that sought-after
combination of riparian
zone – and sheltered
woodland.

OPPOSITE:
Pollution problems
Even in reserves, modern
life can have an impact
on wildlife. Car exhaust
fumes, industrial gases,
suspended particulate
matter, and contaminated
water from agriculture,
mining and logging are
an ever-present problem
which can contaminate
the food-chain and
potentially poison
wildlife. Ranthambore's
relatively isolated location
helps reduce those risks.

FOLLOWING PAGES:
**Cross-country
cooperation**
The 13 tiger range
countries have committed
to doubling their big cat
populations by 2022 –
the next Chinese year of
the tiger. The project,
called Tx2, focusses – for
the first time – not on
saving one sub-species
in one country, but
working across borders
to maintain habitats and
wildlife corridors.

Money-spinners

Tiger cubs are a popular attraction at Ranthambore National Park, and, if the value of nature must be measured in terms of money, then these magnificent cats certainly earn their keep. It's been estimated that each tiger puts £1.8 million (US$2.19 million) into India's economy every year.

FOLLOWING PAGES:
Fighting fear

Pench National Park in India is thought to have inspired Kipling's *Jungle Book* stories. Kipling's Shere Khan was a would-be man-eater, and tales like this have exaggerated tigers' danger to humans and hindered conservation efforts. Education is therefore as important as habitat in saving these rare cats.

BELOW:

Perfect for baby

As every new mother knows, getting the nursery just right is vital. For this pregnant tigress, any den needs to be in a secluded spot, hidden from other predators, and protected from the extremes of the weather. Caves, tree hollows or thick grass are perfect.

OPPOSITE:

Going, going, gone?

In 2007, conservationist George Schaller, talking about India, summed up the problem facing all tiger range countries by saying each nation must decide if they want to keep their natural heritage for future generations to enjoy, 'a heritage more important than the cultural one … because once destroyed it cannot be replaced'.

Going home

The Tiger Summit in 2010, proposed that tiger range countries that didn't then have breeding tiger populations (Cambodia, Lao, Myanmar and Vietnam) but still had suitable habitats should commit to the reintroduction of tigers in those areas. This could see tigers from successful reserves relocated back to their traditional ecosystems.

Getting to know you

When a female tiger is ready to mate, she signals by scent-marking trees. Where territories are large or tiger densities low, the female (right) may have to go in search of a male. It can take several encounters and lots of grooming and nuzzling before they trust each other enough to mate.

Old ways, new ways

In the past, Siberian tigers were revered by local tribes. The Tungus peoples considered them guardians of the forest and addressed them as 'Grandfather'. When Russia and China claimed these regions, they hunted the tigers, who stood between them and the land's 'valuable' resources (ginseng, sable fur and musk ox oil).

OPPOSITE:

Snow cats

The Siberian tiger's
primary habitat is
the Russian taiga –
stretching from temperate
woodlands to snowy
forests 914m (3000ft)
above sea level. Here,
winter temperatures
may fall as low as -40°C
(-40°F) and snow, up to
50cm (20in) deep, lies on
the ground for a quarter
of the year.

ABOVE:

What's in a name?

The spectacularly
beautiful region known
as the Russian far east
encompasses much of the
Amur River delta. This
area has long been one of
the last strong-holds of
Russia's tiger population
and the reason why
Siberian tigers are also
known as Amur tigers.

Little sticks

In Russian, taiga means 'land of little sticks'. Trees in the taiga are shorter than those in other regions due to the reduced growing season and poor-quality soil. The further north you go, the sparser the trees become and the landscape changes into tundra, which comes from another Russian world meaning 'treeless marshy area'.

PREVIOUS PAGE LEFT:
A happy ending?
Every picture tells a
story and, for this tiger,
it has a happy ending.
Once, Meow was kept
at a gas station, where
customers paid to have
photos taken with her.
Rescued by the Kao Look
Chang Wildlife Centre,
Thailand, she's now safe
in an environment where
exploited animals get
a taste of life as it was
meant to be.

PREVIOUS PAGE RIGHT:
Hit or miss
Tigers are usually
nocturnal hunters, using
their superb night-sight
and their natural stripy
camouflage to blend into
the environment. Even
with these advantages,
conservationist George
Schaller's studies of
Bengal tigers revealed that
as few as one in every 20
hunts resulted in a kill.

RIGHT:
Scented messages
Urine sprayed on trees,
on the edges of a tiger's
habitat, has two purposes.
It warns other tigers to stay
away, while also advertising
the tiger's presence to
potential mates.

ABOVE:
Selectively bred
Most white Bengal tigers
are the product of selective
breeding and are kept
as attractions in wildlife
parks. Naturally, they
would share the same
habitat as their golden-
coloured relatives but they
are extremely rare, if not
extinct, in the wild. The
last confirmed sighting
was fifty years ago.

RIGHT:
Cautious climbers
Young tigers are light and
agile enough to make the
most of the trees in their
environment, occasionally
chasing monkeys or
other small mammals
and reptiles up, into the
foliage. Hungry adults,
especially mothers with
cubs to feed, may do the
same, but they're cautious
climbers.

LEFT:
Ambush hunters
Tigers can reach a top sprint speed of 65kmh (40mph). They can only sustain such an incredible spurt over short distances and in wooded areas it's virtually impossible for them to pursue prey through trees. So they rely on the element of surprise, ambushing their prey.

FOLLOWING PAGES:
Popping out for a bite
Tigers spend much of their day resting and grooming themselves. If food is scarce, or if a tigress has cubs to feed, they will venture out during the day, favouring cloudy overcast conditions. While females hunt within their own territories, males tend to range further afield.

Protect and save

In the 13 tiger range countries are nine major water courses, which provide water for 830 million people across India, Indonesia, Malaysia and Thailand. Protecting tiger forests means more clean water for everyone, because forests limit the amount of sediment reaching rivers and reservoirs.

Combatting climate change
Whether it's in Russia (top) or India (bottom), safeguarding tiger habitats will also have a positive impact on efforts to combat climate change. Trees are critical for carbon storage. One tree can absorb as much as 22kg (48lbs) of carbon dioxide every year.

As they grow, trees soak up and store CO_2, which is one of the major greenhouse gasses. It's been estimated that the Amur tiger reserves in the Russian far east can absorb 130,000 tons of carbon per year, which is the equivalent to the carbon dioxide produced by 25,000 cars.

ABOVE:
Good neighbours
Forests regulate water flow. The presence of trees means that water enters the ground more slowly. Cut down tress and you get flash floods. Soil is washed away – and sometimes homes and lives are lost in the process. Once again, protecting tigers has positive repercussions for everyone who lives with these magnificent beasts as neighbours.

Help for all?

Asian elephants, Sumatran rhinos, and Sumatran orangutangs are all critically endangered and all three species can be found in areas traditionally occupied by tigers. That means protecting tiger habitats may also protect some of the Earth's other increasingly rare species.

Meet the locals

The Jim Corbett National Park is a forested sanctuary in India's northern Uttarakhand state. With vast swathes of thick jungle, the Ramganga River and diverse prey, it's ideal for tigers. Visitors are only allowed into certain areas and some tigers have become so popular they're known by name to guides and visitors.

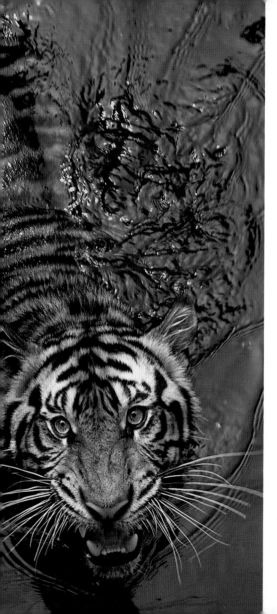

An oxymoron?

Ecotourism is a rapidly growing area, which aims to combine the thrill of travelling with educating tourists, and providing funds for ongoing environmental projects. However, many people view ecotourism as an oxymoron – air travel, and the pollution involved in bringing people to isolated areas, may cause more problems than it solves.

Getting from A to B

The Sumatran tigers' preferred habitat is areas of steep, uncultivated forest. They need natural, unbroken forest corridors in which to form territories. But where plantations encroach on their range, they will avoid humans, using watercourses and forested trackways, covered with thick plant cover to move about in safety.

Hunting for Food

Forty thousand years ago, when our ancient ancestors sat huddled around their fires, telling tales to chase away the night-fears, the stories that grabbed their imaginations the most were about the natural world. Look at their cave paintings and you'll see a rich tapestry of art filled with images of everyday life, and the animals they shared their world with.

India's prehistoric rock shelters in Bhimbetka show pictures of bison, crocodiles, lions and tigers. Hunting scenes are common and it seems our forebears were just as fascinated by the interplay between prey and predator as we documentary lovers are today. Because, while tigers are beautiful, there's no doubt that much of their appeal is the way that they dominate and shape their environment. Tigers are apex predators. They sit at the top of their food chain, and rarely compete for resources with other animals. Yet they're far more than just powerful killers.

Tigers are an umbrella species: protecting them has the knock-on effect of protecting other species in the same region. And, because tigers need a lot of space to roam, their presence ensures protection for their habitats in which they live. There's increasing evidence, too, that predators like tigers are vital to a well-functioning ecosystem. They regulate herbivore populations, ensuring that their numbers don't grow unchecked.

OPPOSITE:

The hunter, hunted
The biggest threat to these great hunters is another hunter: humans. Tigers are hunted for their skins and body parts, which are collected as trophies and used in traditional medicines. Traffic International, a wildlife monitoring agency, says the body parts of 1069 tigers – almost half the world's tiger population – have been seized by anti-poaching officials in the last decade.

Raw power

Tigers are fast, but they can't maintain their speed for long. Most of the animals that tigers prey on can, in fact, easily outrun them. So attacks are usually made from an ambush position – from the side or behind – with the tiger using its powerful forelegs to grasp and pull prey down.

ABOVE:
Stealthy stalker
While hunting, a tiger uses its environment for concealment. Gradually edging closer and closer to its target, at about six to nine metres away (20–30 ft), it's ready to pounce. It's thanks to this behaviour that a group of tigers is known as an ambush or streak.

OPPOSITE:
Slowly does it
When stalking prey, patience is everything. Tigers can spend 20 minutes or more approaching their target. If their prey is spooked and bolts before the tiger can pounce, then the attack is likely to be unsuccessful and the big cat will go hungry.

A waiting game

When prey is in sight, the tiger freezes, then slowly lies down in the long grass. As long as it remains motionless, and there's no breeze to carry its scent, it stays hidden, waiting for its prey to lower its guard, or for a young or weak animal to get separated from the herd.

Clever camouflage

A tiger's stripes break up its body shape, making it hard to spot in long grass and thick undergrowth. As tigers usually hunt at night, or in low-light conditions, the stripes can easily be mistaken for shadows and ignored by unwary animals until it's too late.

Eliminating the competition

In 2014, researchers found evidence of a rare cat-on-cat attack: a Siberian tiger killing an Eurasian lynx. It was clear from the carcass that the tiger hadn't killed the wild cat because it was hungry. Instead it's believed that it was killed to eliminate the competition for food.

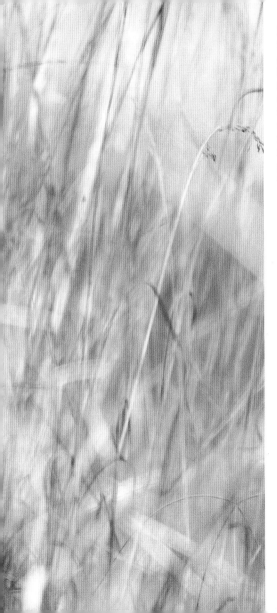

In living colour

In daylight, tigers can
see as well as the average
human. Many predators
are colour blind and, until
recently, it was thought
that a tiger's eyes were
designed to detect detail
rather than colours. It's
now believed they can see
greens, blues, yellows and
shades of grey.

Familiar felines

If you have a pet cat, then you'll have seen how it stretches when it wakes, pushing out its forelegs and arching its back so that its rear rises up. This remarkably similar stance from a Bengal tiger shouldn't be so surprising. Tigers and domestic cats share around 95 per cent of the same DNA.

ABOVE:

Poetry in motion

Cats walk by lifting both legs on the same side of the body together. A tiger, in a full run, changes its gait so that only one paw touches the ground at a time. This Siberian tiger has been caught between steps, so it appears to have left the ground completely.

OPPOSITE:

Built for speed

Tigers have a relatively small collarbone for such a large mammal. A small collarbone allows for a wider, unrestricted range of movement. Meaning the tiger can move its shoulders much further when running, covering longer distances with each stride.

ABOVE:
Tiger tails
Tigers have long tails
– sometimes growing
up to a metre (3.3ft) in
length. Such a notable
feature isn't just for show.
Although they do use
it to communicate with
other tigers, it's primarily
for balance, and acts
like a rudder to help the
cat make sharp turns at
speed.

OPPOSITE:
Brace for impact
A tiger's paws and claws
are one of its greatest
weapons during the hunt.
Strong, and powerful
enough to kill prey with
a swipe, the bones in
their paws also have
cord-like ligaments to
buffer them from the
impact of hitting prey at
a full run.

FOLLOWING PAGES:
The eyes have it
Domestic cats have
pupils that are vertical
slits. Tigers and lions
have round pupils. The
theory is that vertical
pupils maximise the
ability of small animals
to judge distances, while
round pupils work better
for larger predators.
Horizontal pupils are
more common in prey
species, as they expand
the field of view.

OPPOSITE:
Nature's naturalist
In *The Face of the Tiger*, the ecologist Charles McDougal recorded his observations about tigers saying: 'The tiger is a first-class naturalist and knows the seasonal and activity patterns of various prey animals – where they may be found, and when they will be feeding or resting'.

ABOVE:
Swift and deadly
Tigers may stake out water holes, hiding in the grass until prey appears. Chasing its terrified target into the deep water, the big cat will pounce, pushing its prey underwater. Prey will drown or suffocate as the tiger's canines close on its neck, breaking the spinal cord, piercing the windpipe, or severing the jugular.

ABOVE:
A varied diet
Tigers are opportunists, hunting whatever comes their way. What they eat depends on their location. In Siberia, around half of their diet consists of wild boar. In India, sambar deer account for almost sixty per cent of a Bengal tiger's prey. Tigers will eat rodents, reptiles and fish if larger prey species aren't available.

OPPOSITE:
Food storage
Leopards will haul a kill up a tree to keep it safe from other predators. Tigers have a less energetic way of ensuring that their food isn't stolen: they cover it in soil and leaf-litter. They'll return to the same kill over and over until it is consumed.

Energy saving

A tiger needs to make a kill at least once a week. A large animal may feed the tiger for two or three days, then it rests, before hunting again. A huge amount of energy is expanded in a hunt, so it's more efficient to kill one sizeable animal than lots of small ones.

LEFT:
Amazing adaptations
Tiger stripes are designed
to break up the cat's
outline in undergrowth.
This is little use in the
snowy north, so Siberian
tigers have the fewest and
the palest stripes of any
sub-species. Their large
paws act like snowshoes,
to stop them sinking in
the snow.

FOLLOWING PAGES:
**Watch out – trouble's
about!**
These chital deer are
one of the Bengal tigers'
favourite prey animals.
However, they aren't
about to make it easy
for this hungry tigress.
Standing, facing towards
the predator, the whole
herd is alert to her every
move and they can be very
vocal when alarmed.

Looking for trouble

These deer would appear to be easy pickings for the tiger, but their alert stance – ears pricked, sniffing the air – tells us that they're well aware danger is near. If the tiger makes one wrong move, the deer will scatter and hours of watching and waiting will be wasted.

FOLLOWING PAGES:

Food favourites

Sambars are the world's largest deer, and a favourite food of the Bengal tiger. They've even been known to mimic the deer's call to lure them close enough to ambush. One swipe from a tiger's paw is enough to seriously injure a stag weighing 150kg (330lb).

Good neighbours

A sambar deer's alarm call is a throaty, booming 'pooking' sound, sometimes accompanied by agitated foot stamping. Different animals have been known to cooperate when a tiger is in the area, sending out loud, piercing calls to warn not just members of their own species, but others too.

OPPOSITE:
Pressure-sensitive
Tigers have fewer teeth
than many carnivores
but, what they lack in
number, they make up
for in size and sensitivity.
These huge canine teeth
contain pressure-sensitive
nerves allowing the tiger
to judge exactly how hard
to bite as it clamps onto
its prey's throat.

ABOVE:
Mother love
In this incredible image, a
mother sloth bear charges
a large male tiger who's
threatening her cub. The
bear stands on her hind
legs and swipes at the
cat with her long curved
paws, teeth bared. While
tigers do prey on bears,
this determined mother
left the would-be attacker
seriously injured.

FOLLOWING PAGES:
Strength in numbers
Different species have
different ways of coping
with predators. These
chital deer flee in groups,
making it harder for a
lone hunter to separate
one animal from the herd.
Typically, when grazing,
they'll stay close to
cover and dive back into
the undergrowth when
danger strikes.

LEFT:
Using your head
Usually a tiger will only become a man-eater if they're too old or injured to hunt any other prey. As tigers attack from behind – relying in stealth and surprise – farm workers in India have been known to wear masks on the back of their heads, so that the tiger doesn't sneak up on them from behind.

FOLLOWING PAGES:
Cat fights
It's not unusual for tigers in India to share their range with other predators, such as leopards and the dhole – a relative of the wolf. These don't compete directly with the larger cat for food, but when it's hungry, they may still end up on the menu.

OPPOSITE:

A toothy problem

Using their superior size and strength to knock prey down, the tiger's killing bite is made to the back of the neck. Once latched on to its target, it stays locked on until its prey is dead. However, hitting prey at speed risks broken teeth, which can spell starvation and early death for the tiger.

ABOVE:

Table manners

Wildlife experts find it easy to identify a tiger kill because of the very particular way that the cat dissects its food. They start at the animal's hindquarters, carefully removing the parts they don't want – pulling out the intestines and rumen – then use their rough tongue to clean the bones.

FOLLOWING PAGES:

Going it alone

On open grasslands there are few places to hide, so lions need the help of a pack to trap and encircle prey. Tigers live in habitats where heavy foliage allows them to ambush prey without help. It's likely that this small group of Bengal tigers is a mother and her cubs.

Meat and veg

A herbivore's intestines are 10–12 times the length of their body, because it takes a lot time to break down and absorb plant matter. Meat is easier to break down so carnivores only need intestines that are three to six times their body length.

FOLLOWING PAGES:

First come, first served

In many predator groupings, the alpha – the biggest and strongest male – eats first. Tigers will share a kill, letting their mate and their cubs eat first. Occasionally, tigers will share a kill with other males, but the cat whose kill it was will eat first, even if he's not the dominant male.

ABOVE:

Big appetites

Predators often gorge on meat when it's available, eating huge amounts in one sitting. It's not unusual for a tiger to consume almost 2kg (4lb) of meat at a time. The larger Bengal tigers eat an average of 6kg (13lb) a day. A mother with two, fast growing cubs needs about 9kg (20lb).

OPPOSITE:

Learning the ropes

At about eight weeks old, mothers begin teaching their cubs how to be self-sufficient. They'll accompany her on the hunt and she even may bring them an injured animal to kill. They're able to hunt for themselves at about 18 months of age, but stay with mum until they are about 2.5 years old.

Tiger Cubs

Adult tigers are an awe-inspiring spectacle – combining strength, power, and agility. However, tiger cubs are just as vulnerable as any newborn, and their early months are fraught with danger. Born blind and totally reliant on their mother, the cubs that survive their first year face a life full of deadly challenges. Cubs are playful and adventurous by nature but, in the wild, there are threats everywhere.

There's also the very real risks posed by male tigers, who may kill cubs in order to bring the female into oestrus, so that she's ready to mate with him. Only around half of all young tigers reach adulthood and, for the tigress, the work of keeping them safe starts immediately.

Tigers can breed any time of the year, but favour the cooler months, between November and April. Choosing a den spot that's secluded and sheltered, mum rarely strays far from home during these perilous first few months. Around 70 per cent of her time is spent caring for her cubs and, if she feels like the family is in danger at any point, she'll quickly relocate – gently carrying them in her mouth to safety. The cubs drink mum's milk until they're weaned at around six to eight weeks, when she'll introduce them to solid food and begin taking them hunting. Following mum's lead, they quickly develop the skills and techniques that will serve them into adulthood.

OPPOSITE:
Family bonds
The tigress gives birth between 13 and 15 weeks after conception. Litters of up to seven have been recorded, but two to four is more usual. Strong familial bonds are established through regular physical contact, with cubs nuzzling against their mother in the same way that domestic cats rub against their owners.

Blue-eyed babes

Cubs open their eyes between six and 12 days after they're born. All tigers are born with blue eyes. In most sub-species these gradually change to a more striking yellow-ochre colour as they grow. White Bengal tigers retain their blue eyes into adulthood.

ABOVE:
Working mums
These cubs may suckle at their mother's teat for up to 45 minutes at a time. Because mum also needs to eat, she'll tuck her cubs safely away in the den then hunt through the night, returning home to feed her hungry family again in the morning.

OPPOSITE:
Difficult choices
If food is plentiful, then their mother will ensure that each cub has enough to eat. If food is scarce, then it becomes a survival-of-the-fittest situation and the mother will give the dominant cub more food, as he or she is the strongest of the litter and so the most likely to survive.

FOLLOWING PAGES:
A helping paw
Males usually mate with females whose territories overlaps theirs. Once they've mated, they generally go their separate ways, but males have been known to play, feed and even help raise cubs. This makes sense for the male, as it increases the likelihood of the cubs surviving and, therefore, his genes being passed on to the next generation.

Tiger territory
Successfully raising a
family needs just the
right habitat. India's
Bandhavgarh National
Park (shown here) boasts
mixed forests, wide
valleys and streams,
making it ideal tiger
territory. The reserve has
the one of the densest
populations of tigers in
India and a high rate of
tiger cubs surviving to
adulthood.

OPPOSITE:

Cat communication

Tigers cubs make adorable yawning yowls. It takes around 20 months until the youngsters are physically developed enough to replicate the adult's famous roar. This roar isn't just terrifying, it's incredibly complex and includes low-frequency 'infrasound' vocalizations, which can travel long distances, cutting through dense forests – even mountains.

ABOVE:

Survival plans

Over the years, public perceptions and expectations of zoos have changed. Today, we trust good zoos to do more than provide a place to view animals. They should also play a role in helping endangered species, through breeding programs. However, captive breeding is pointless unless the causes of species decline are tackled, and some zoos are also actively engaged in advocacy and education about habitat loss and climate change. In nature only one in every 10,000 Bengal tigers are born white and, apart from generating interest in tigers, breeders of white tigers don't contribute to any species survival plans.

Mum knows best

In captivity, tiger cubs don't learn the skills needed to survive in the wild. Mum doesn't need to hunt, so she never passes on those vital life lessons. In the wild, things are very different. Cubs continually watch and mimic mum's behaviour and, by two years of age, they're ready to head off on their own.

FOLLOWING PAGES:

Late bloomers

Tigresses are scrupulous about cleanliness, and food will never be left inside the den as it risks disease. Mum will start to leave meat outside the den for her cubs as soon as they've developed their canine teeth. In captivity Siberian tiger cubs (shown) are late bloomers and don't start eating meat until they're around 10 weeks old.

Tiny explorers

Adult tigers spend most of their time asleep – and they'll sleep anywhere they feel safe, even next to a fresh kill. Their cubs are more active and, as they grow in strength and confidence, they become increasingly inquisitive about their environment, which can lead them into trouble.

Spring babies

Depending on the sub-species, a tiger cub weighs an average of 1kg (2.2lb) at birth. Most cubs are born between March and June and a tigress generally gives birth once every two years, producing cubs for the first time when she's four or five years old.

Teething tigers

Like humans, tigers are born without teeth. The first, needle-sharp milk teeth appear after a few days. At around six months old, a set of new, permanent adult teeth will begin to push these milk teeth out. These grow throughout the tiger's life, and it's possible to age a tiger by the size of its teeth.

Top cat

In every tiger litter there's always one dominant cub. It's this top cat who leads the others in play activities and may even dictate when they eat and sleep. This can be a male or female, but is more likely to be a male, because they're bigger.

LEFT:
Gentle jaws
A tiger's jaws are designed to crush bone, but it's not indiscriminate power. The same jaws can also be used to gently lift a cub. When the mother clamps down on her cub's neck, its reflexes kick in. It goes limp to avoid injury, back legs curled slightly so not to catch on undergrowth.

ABOVE:
Special saliva
Big cats are as fastidious about hygiene as their domestic counterparts. The tigress's sandpaper-like tongue makes the ideal tool for keeping her cubs clean. Her saliva is also antiseptic, which means that she is able to disinfect any cuts or wounds with a simple flick of the tongue.

Swimming tests
In the wild, tigers learn
to swim by following
mum into the shallows,
gradually gaining enough
confidence to go into
deeper water. In zoos,
tiger enclosures are often
surrounded by water
and, after some tragic
drownings, some zoos
now give 'swimming tests'
to cubs before they can
join their parents in the
adult enclosure.

Chilling in the pool
Large animals have more
surface area to heat up,
and once they're hot it's
hard for them to cool
down. As the largest
member of the genus
Panthera, the tiger is more
likely to overheat than its
smaller cousins. Which is
probably why it's one of
the few cats who like to
cool down with a swim.

Winners and losers

From an early age, tiger cubs play-fight, learning to stalk, swipe, and pounce. Adults will fight to win or defend territory, but not every encounter ends in blood. Sometimes intimidation works and, if the weaker tiger submits to the threats of the more dominant one, then the winner may tolerate the loser sharing its range.

Hide-and-seek
Tiger cubs love playing
hide-and-seek games
with mum but, like
all tiger-play, there's a
serious purpose behind
the fun. She hides in the
undergrowth, while the
cubs practice pouncing
on her tail – preparing
them for a time when
they'll have to hunt for
themselves.

A double-edged sword
The tigress will play-fight
with her cubs, helping
them develop their attack
and defensive skills.
For mum, this can be a
double-edged sword. Her
daughters will establish
ranges close to hers and,
one day, may use the same
fighting skills to challenge
her for her territory.

LEFT:

Caring for the cubs

There's more going on in this photo than mother–cub bonding. It's thought that the tigress stimulates her cubs' circulation and digestion by licking them. Cats only sweat through their paws, so licking her cubs may also be a way of ensuring that they don't overheat.

FOLLOWING PAGES:

Smell you later

Tigers don't rely on smell for hunting, but it plays an important role in cat communication.
Cubs recognize mum, and each other, by scent. Tigers, like other cats, have a particular facial expression they use when sniffing – called flehmen. Wrinkling the nose and letting the tongue hang out directs scents to the Jacobson's organ, behind the front incisors, which detects odour particles.

ABOVE:

**Black, white
and blue too**

In addition to white
tigers, black and blue
tigers are also known.
Black tigers have a dark
grey or black coat with
lighter grey or pale
tan stripes. Blue tigers
have a bluish-grey coat
with charcoal-grey
coloured stripes. Both
are incredibly rare, if not
extinct, in the wild.

OPPOSITE:

From head to tail

From swift, running
turns, to standing
jumps, it's the tiger's
backbone that allows it
to make such acrobatic
movements. Its extremely
flexible spine contains 30
vertebrae (humans only
have 25) and it extends
from the base of the skull
to the tip of the tail.

FOLLOWING PAGES:

Sheathe your weapons

During this mock fight,
the cubs' claws are
sheathed. Leopards, lions,
jaguars and tigers can all
fully retract their claws
when they're not in use.
Cheetahs are so fast they
need their claws for extra
traction while running,
and they are one of the
few wild cats whose claws
are only semi-retractable.

OPPOSITE:

Sexual dimorphism

Male tigers are longer in the body and around 1.7 times heavier than their female counterparts – and this begins to become apparent once tigers are weaned. The females also have a shorter head, shorter canines and smaller paws. This difference in size between the sexes is known as sexual dimorphism.

ABOVE:

King of the beasts

In Chinese legend, the markings on a tiger's forehead are said to represent the pictogram for '*guó wáng*' which means king. At the start of the year of the tiger, it's traditional for children to have this character painted on their foreheads in wine, to promote strength and health.

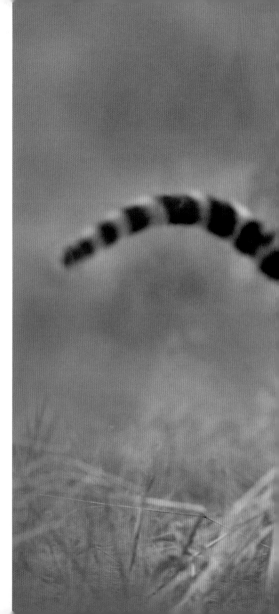

RIGHT:
Tail tales
When it's not being used to help the tiger keep its balance, these big cats use their tails to signal their emotions. A relaxed tiger has their tail hanging loosely down. Anger or irritation is shown by a rapidly flicking tail.

FOLLOWING PAGE LEFT:
Leaving your mark
Tigers mark the boundaries of their territory using urine, faeces, and scent glands, alongside visual signals such as claw-marks on the ground or on trees. This Siberian tiger cub is still too young to have its own territory but may be practicing for a time when it does.

FOLLOWING PAGE RIGHT:
Moving home
Tiger mothers are hugely protective of their little charges. Little is known about exactly what happens inside a wild tiger's den, because she goes to great lengths to keep her cubs hidden. If she believes the den has been disturbed, then she will immediately move her family to a new location.

Arboreal ancestors

Proailurus was a distant relative of today's big cats that lived around 25 million years ago. It's believed that the species was partially arboreal, spending much of its time in the tree-tops. All of today's cats retain some tree climbing abilities but adult tigers tend to be happier on the ground.

OPPOSITE:
A struggle to survive
White tigers struggle to survive in the wild. The same dramatic colouring that makes them so popular as attractions also hinders their ability to hide and stalk their prey undetected. Without the more usual orange and black camouflage, white tigers are likely to starve to death.

BELOW:
Overprotective mothers
In India, the country's total possible tiger habitat is around 300,000 sq km (115,830 sq mi) but tigers are squeezed into about 10 per cent of that space. This means that encounters with humans are increasingly likely and tiger attacks, often by mothers protecting their cubs, are on the rise.

FOLLOWING PAGES:
Preparing for motherhood
A female tiger comes into oestrus – meaning that she's likely to conceive – once every three to nine weeks, depending on the sub-species and geographical location. During oestrus, the male and female may mate several times every hour which increases the chances of her getting pregnant.

Stay-at-home mums

Due to their need to stay close to their cubs, female tigers generally only hunt within their own territory. Males – with no cubs to care for – are free to travel further afield and will range far and wide, even venturing into other tigers' territories.

Natural hunters?

It's been said that tiger cubs recognize prey species instinctively. In one experiment, young tigers who had never seen a real deer attacked a model of a deer, which had been seeded with deer urine. However, it still takes years of training to turn instincts into skill.

A difficult transition

At around two years old, tigers begin independent lives. This is a dangerous time for young tigers, with no territories of their own and no adult protection. The mortality rate is as high as 35 per cent for these untested teens, the lessons they've learnt from mum can be the difference between life and death.

Picture Credits